Among Other Things

BEN HOLDEN-CROWTHER

MMXVII

HAM THE CHIMP

Ham (1956 – January 19, 1983), also known as Ham the Chimp and Ham the Astrochimp, was a chimpanzee and the first hominid launched into space, on 31 January 1961, as part of the US space program.

GEOSAURUS AND BELODON

Geosaurus was a sea creature which lived during the Triassic period. Belodon was a crocodile-like dinosaur.

MARTIAN MEMORIAL

This image taken by the panoramic camera aboard the
Mars Exploration Rover Opportunity shows the rover's
empty lander, the Challenger Memorial Station, at
Meridiani Planum, Mars.

LANDSCAPE WITH BIRDS (1628)

Orient the image. Can you spot the extinct animal?

DODO

This bird was endemic to Mauritius.

EARTHRISE

Taken by Apollo 8 crewmember Bill Anders on
December 24, 1968, at mission time 075:49:07.

THE INTERNATIONAL SPACE STATION (ISS)

Visible with the naked eye from Earth. Work on this
habitable satellite began in 1998.

MOUNT PAVLOF

One of the most active American volcanoes. It erupted
twice in 2014 as well as in 1980, 1981, 1983, 1986–
1988, 1996–1997, 2007, 2013.

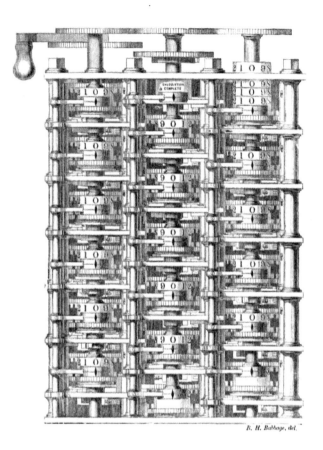

DIFFERENCE ENGINE

A mechanical computer designed by Charles Babbage. It was assembled in 1833.

THE GLIDING MAN

Otto Lilienthal performing one of his gliding
experiments around 1895.

LONDON UNDERGROUND

The world's first underground railway, the Metropolitan
Railway, which opened in 1863, is now part of the
Circle, Hammersmith & City and Metropolitan lines.

LONDON BUSES

Buses have been used on the streets of London since 1829, when George Shillibeer started operating his horse drawn omnibus service from Paddington to the city.

NEBULA

Made up of gas and dust, the pillar shown resides in a
stellar nursery called the Carina Nebula, located 7500
light-years away in the southern constellation of Carina.

WONDERS OF THE WORLD

The Seven Wonders of the Ancient World:
Great Pyramid of Giza, Hanging Gardens of Babylon,
Temple of Artemis at Ephesus, Statue of Zeus at
Olympia, Mausoleum at Halicarnassus, Colossus of
Rhodes, and the Lighthouse of Alexandria.

KING TUT

The 1922 discovery by Howard Carter and Lord Carnarvon of Tutankhamun's nearly intact tomb received worldwide press coverage. In 2010, the results of DNA tests confirmed that he was the son of Akhenaten.

DRILLING FOR LIFE

This image shows Curiosity's robotic arm showing drill in place, with Yellowknife Bay and Mount Sharp in the background.

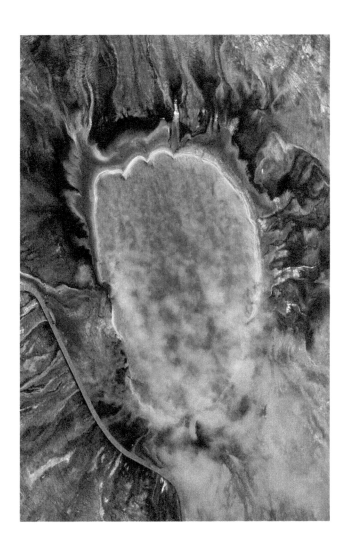

GRAND PRISMATIC SPRING

Orient the image to see an aerial view of the Grand Prismatic Spring at Yellowstone National Park. Mats of orange algae and bacteria surround the main spring.

EMBRYO

Even an embryo this small shows distinct anatomic features, including tail, limb buds, heart, eye cups, cornea/lens, brain, and prominent segmentation into somites.

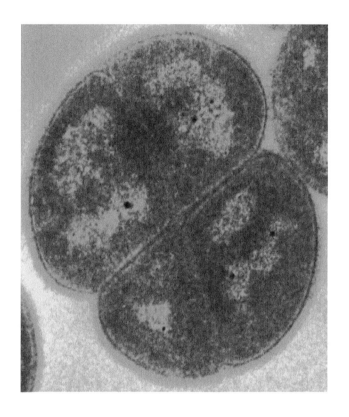

SURVIVAL

Deinococcus radiodurans is an extremophile that can resist extremes of cold, dehydration, vacuum, acidity, and radiation exposure.

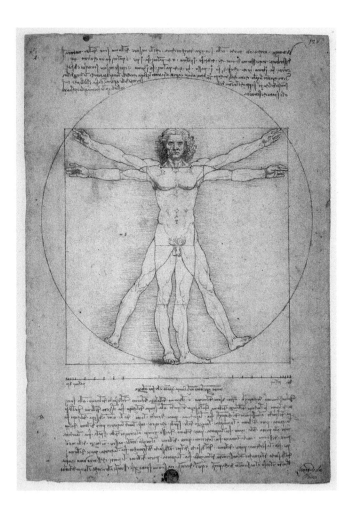

VITRUVIAN MAN

Vitruvian Man, Leonardo da Vinci's image is often used as an implied symbol of the essential symmetry of the human body, and by extension, of the whole universe.

ATOM BOMB

Mushroom cloud above Nagasaki after atomic bombing
on August 9, 1945. The photo was taken from the north
west. On the day of the nuclear strike (August 9, 1945)
the population in Nagasaki was estimated to be 263,000,
made up of 240,000 Japanese residents, 10,000 Korean
residents, 9,000 Japanese soldiers, and 400 POWs.

RELIGION

A view of the *gopuram* at the entrance of the Sri
Mariamman Temple in Singapore.

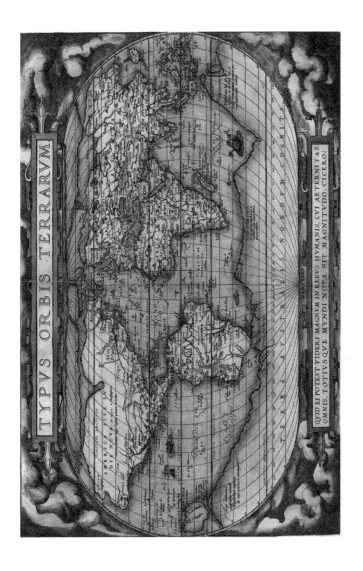

WORLD MAP

Orient the image to see a world map devised by
Abraham Ortelius in 1570.

MEN ON THE MOON

In December 1972, Apollo 17 astronauts Eugene Cernan
and Harrison Schmitt spent almost 80 hours on the
Moon in the Taurus–Littrow valley.

THE HINDENBURG DISASTER

This photo was taken of the crash of the Hindenburg
airship in Lakehurst, New Jersey on May 6, 1937.

BEOWULF

Beowulf is an Old English epic poem consisting of 3182
alliterative lines. It is believed to be the oldest surviving
long poem in Old English.

SEXTANT

Astronomy became much more accurate after Tycho
Brahe devised his scientific instruments for measuring
angles between two celestial bodies.